Primary Geography

Pupil Book 1 and 2 World around us

Stephen Scoffham | Colin Bridge

Earth in space
Earth, sun and moon	4-5
The planets	6-7
Day and night	8-9
Land and sea	10-11

Planet Earth
A living planet	12-13
The shape of the land	14-15
Volcanoes	16-17
World wonders	18-19

Weather and seasons
Experiencing the weather	20-21
Different types of weather	22-23
Extreme weather	24-25
The seasons	26-27
Going round the sun	28-29

Local areas
Shelter	30-31
Houses around the world	32-33
Living in a village	34-35
Exploring local streets	36-37
Under your feet	38-39

Maps and plans
Maps and stories	40-41
Treasure island	42-43
Different plans	44-45
The view from above	46-47

The UK
| UK countries | 48-49 |
| UK mountains and rivers | 50-51 |

Different environments
Living in the arctic	52-53
Living in the rainforest	54-55
Living in the desert	56-57
Animals around the world	58-59

World maps
| World continents | 60-61 |
| World countries | 62-63 |

The world around us

Geography is about the world we live in. It is about volcanoes, mountains, rivers and deserts. It is about great cities and distant places. It is also about the place where you live. What it is like and how it is changing.

Here are some of the characters who are going to help you learn more about geography.

Baby Squirrel
Tommy Trout
Fran Frog
Max Mole
Alfie Owl
Kate Koala
Polly Parrot
Winnie Worm
Bill and Rita Mouse
Freddie Fox
Billy Bear
Sally Sparrow

Earth, sun and moon

The sun and moon move through the sky above the Earth. The sun brings us warmth. The moon marks out the months.

Baby Squirrel's mistake

Baby Squirrel just couldn't sleep. 'Lie still and you will soon be asleep,' said her mummy.

In a little while Baby Squirrel came downstairs again. 'It's no use,' she said, 'I just can't sleep.'

'Close your eyes and try again,' said mummy.

'But I can't close my eyes, said Baby Squirrel. 'Freddie Fox told me that if I close my eyes the big white ball in the sky will fall on me!'

'Freddie Fox is tricking you,' laughed mummy. 'The big white ball is the moon. It has been in the sky since time began.'

What was Baby Squirrel's mistake?

Talking
- What are the differences between the Earth, sun and moon?

The planets

The Earth is a great ball of rock spinning in space. It is one of eight planets which go round the sun.

A journey in space

The Darin family set off on a journey in their space ship. There was Luke, Alice and Baby Oscar.

Baby Oscar was learning to count. As they left Earth, Alice said, 'How many planets can you see between here and the sun?'

Oscar looked hard. 'One, two,' he said. 'Now add the Earth' said Luke.

'One, two, four,' said Oscar. 'No, no!' laughed Alice. 'One, two, three.'

Soon they passed Mars and Jupiter.

'One, two, three, seven, nine,' said Oscar. 'It's one, two, three, four, five,' chuckled Luke.

Later Saturn, Uranus and Neptune went by.

'One, two, three, four, five, seven, six, eight,' said Oscar.

'Try again,' said Alice, 'but let's have a snack first.'

What planets did the Darin family fly past?

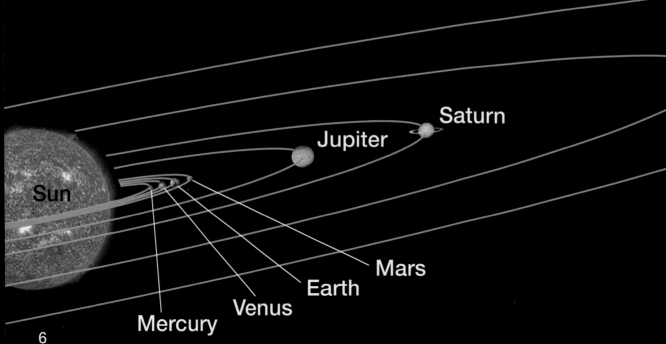

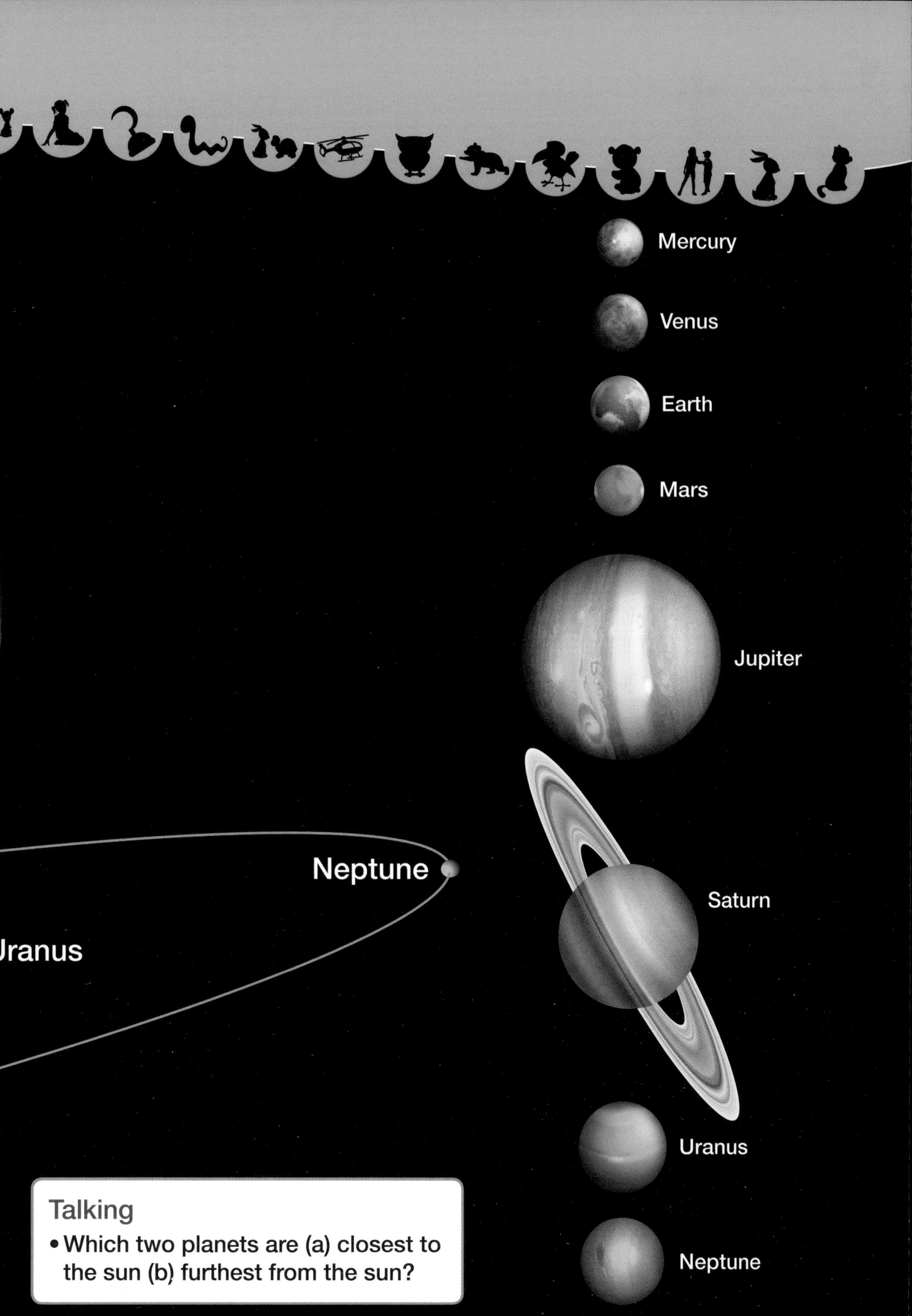

Mercury

Venus

Earth

Mars

Jupiter

Neptune

Uranus

Saturn

Uranus

Neptune

Talking
- Which two planets are (a) closest to the sun (b) furthest from the sun?

Day and night

As the Earth spins in space we get day and night. The sun brings us daylight. In the night we see the moon and the stars.

Flying in the dark

Sally Sparrow was going home from school. 'It will be teatime, bath and bed again,' she thought. 'Every day is the same.'

Along came Brian Bat. 'Hello,' he said, 'Another great night ahead. It's my best time.'

'Do you stay up all night? asked Sally. 'Oh yes!' replied Brian.

That night Sally slipped out of the house and flew around the wood. It was so dark she bumped into trees.

She met Brian again. 'It's horrible at night,' she said, 'I can't see anything.'

'I don't have any problem,' said Brian. 'I have special ears. Sounds help me around.'

'Daytime is best for me,' said Sally.

'Night is an adventure for me,' said Brian.

What was the difference between Sally and Brian?

- North Pole
- Equator
- South

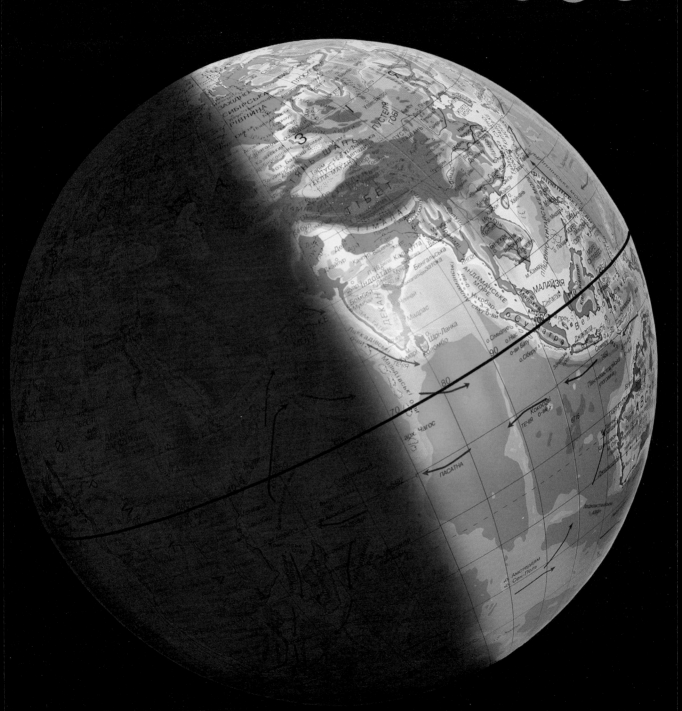

Talking
- What do you like about
  (a) day time (b) night time?

Land and sea

Looking at the Earth from space, astronauts can see land and sea. There is much more sea than land.

Learning to dig

Max Mole lived with his family in a tunnel under the ground. He watched his parents dig new tunnels each day.

'That looks easy,' he said. 'You still have a lot to learn,' said his father.

Next day Max decided to dig his own tunnel. 'This is easy,' he said, and off he went.

Suddenly the soil disappeared. He couldn't dig.

He couldn't even breathe. A strong paw gripped his tail and pulled him back. His father had rescued him.

'You dug into water,' said his father. 'Most of the world is covered in water. You can't dig there. You must learn where to dig, and how to swim!'

What did Max Mole learn?

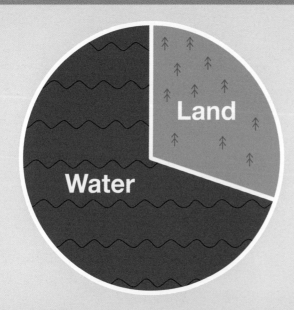

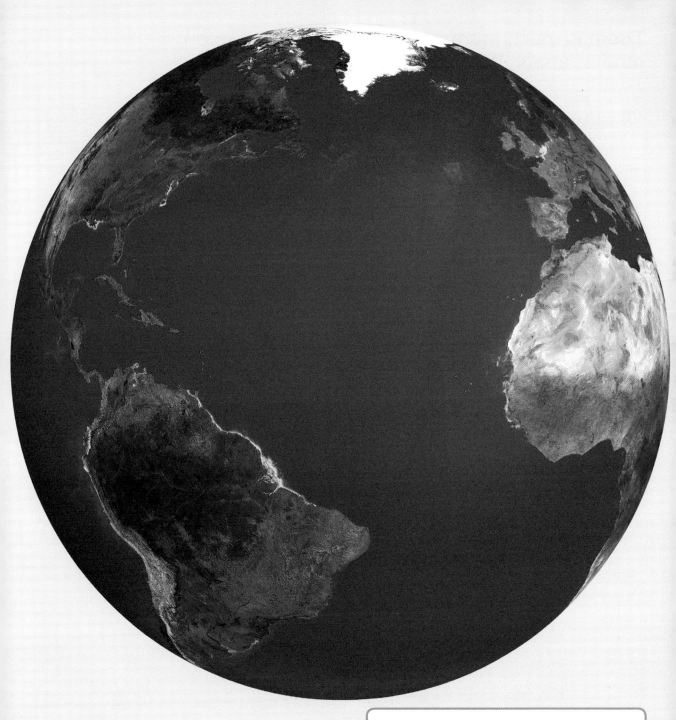

Talking

- Do you think there is more land or sea in the photograph?

A living planet

There is much more water in our world than on other planets. Water brings life to the Earth.

A new life

Little Seed was sad. She had been in the sand in the desert for seven years.

A bird hopped down. 'What are you doing Little Seed?' he said.

'Nothing,' answered Little Seed. 'I can't seem to do anything.'

One day some clouds appeared. They poured rain on the land. 'Oh!' said Little Seed. 'I'm growing! I'm growing!'

Soon she had roots and branches, then flowers and seeds of her own. The water had brought her to life.

Her seeds would wait for the next rainfall to bring the desert to life again.

Why did the rain make Little Seed happy?

Talking
- Why is water important to the plants and creatures in the photographs?

The shape of the land

Mountains are the highest places on the land. Water flows from mountains to the sea in streams and rivers.

A journey to the sea

Tommy Trout lived in a little pool high in the mountains.

'I've heard about the sea,' he thought. 'I'll go and find it.'

Off he went swimming lazily downstream. The water got deeper and the current got stronger.

Suddenly he was flying in the air. 'What was that?' he gasped. 'That was a waterfall,' quacked a duck.

On swam Tommy. There was something shiny in the water. 'Be careful,' croaked a frog, 'That's a fisherman's hook.'

Tommy dashed away down the valley. 'Stay here', said an otter, 'I would like to have you for dinner.' 'No thank you', said Tommy.

'I shall be pleased to get to the sea,' he thought, 'I wonder what will happen there?'

What creatures did Tommy Trout meet on his journey to the sea?

Talking

- What different words describe the river?

Volcanoes

There are hot rocks under the ground. They come to the surface in special mountains called volcanoes.

Surface   Hot rocks
          Core

Strange noises

Billy Bear came running down from his bedroom. 'Mummy, mummy, I am frightened. I can hear a sound like a volcano,' he cried.

'Don't worry,' said mummy. 'It's only daddy snoring.'

Soon Billy came running down again. 'Mummy, mummy, I am frightened. I can hear a sound like a volcano,' he sobbed.

'Don't worry,' said mummy. 'It's only daddy's tummy rumbling.'

Later Billy came running in again. 'Mummy, mummy! Daddy is making a noise like a volcano again.'

'No I'm not,' said daddy waking up. 'It must be a real volcano. Quick everyone into the car. We must go to grandma's until it is safe to come home again.'

Why wasn't it safe for Billy Bear to stay at home?

Talking
- What happens when a volcano erupts?

World wonders

There are many special sights in the world. There are wonders on the land, wonders in the sea and wonders in the sky.

A little bit of magic

Rose was on the beach. She was feeling gloomy. Suddenly she saw a dark green bottle.

Rose pulled out the cork. There was a puff of smoke. 'How can I cheer you up?' said the Genie.

'Make something bright and interesting to see,' said Rose.

Abracadabra! A sparkling white iceberg appeared in the sea. 'That's better,' said Rose, 'but it's very cold. Where's somewhere warmer?'

Abracadabra! She was in a deep cave full of amazing shapes. 'That's better,' said Rose, 'but it's a bit eerie down here.'

Abracadabra! Suddenly the sky was filled with whirling colours. 'That's better, said Rose. 'Now I want some toys.'

'Too late,' boomed the Genie. 'You only had three wishes. Goodbye.'

What wonders did the Genie create for Rose?

Iceberg

18

Cave

Northern Lights

Talking
- What other world wonders can you think of?

Experiencing the weather

The weather never stays the same for long. There is sun, rain and wind. Sometimes it feels hot, sometimes it feels cold.

A nasty surprise

Freddie Fox was out for a walk. It was a bright, sunny day. The birds were singing and the bees were buzzing.

Suddenly he felt a spot of rain. 'Oh! Never mind,' he said. 'It will only be a shower.' He put up his umbrella.

Soon the wind got up. The gentle breeze turned into a howling gale.

It blew into Freddie's umbrella. 'Help,' he cried. 'I'm flying in the air.'

The wind dropped and so did Freddie. Into the village pond.

'An umbrella was meant to keep me dry,' he moaned. 'The weather just isn't fair.'

How was Freddie Fox caught out by the weather?

▲ When it is rainy and sunny we get rainbows.

▼ When it is windy you can fly a kite.

▲ When it snows you can make a snowman.

Talking
- What types of weather do you like most?

Different types of weather

We use special words when we talk about the weather. Sometimes we show what the weather is like using pictures called symbols.

Little Car's holiday

Little Car was looking forward to going on holiday. 'I need a rest,' he said.

He set off in bright sunshine. His paint shone and his engine hummed in the warmth.

Before long some clouds appeared. 'It's nothing,' he said, 'It'll soon change.'

It did! It began to rain. 'It's nothing,' he said, and closed the windows. 'It'll soon change.'

It did! There was thunder and lightning. Water sprayed all over him. Little Car put on his headlights. 'It's nothing,' he said, 'It'll soon change.'

It did! He ran into fog! 'This is awful,' he complained, 'I'm stopped altogether. It'll never change.'

Suddenly the sun came out.

What different types of weather did Little Car experience?

Warm and sunny

Cloud and rain

Sun and cloud

Showers

Fog and mist

Thunder and lightning

Talking
- Talk about all the weather words you can think of.

Extreme weather

Sometimes the weather is wild and exciting. It can be very rainy, windy or hot. This reminds us the weather is very powerful.

A day off school!

Fran Frog hopped slowly home from school. 'I wish I could have a day off,' she grumbled.

That night she was woken up by a crash and a rumble. She looked out of the window. The sky was full of dark clouds. The trees were bent over by the wind.

Suddenly a great flash lit up the sky. It began to rain and rain and rain. In the morning there was water everywhere. The fields were flooded.

'Hurray! No school today,' she thought. She went back to bed.

There was a knock on the door. 'Come on out,' shouted a voice. It was the teacher. She was collecting everyone in her boat.

Poor Fran Frog!

Do you think Fran Frog was frightened by the weather?

Talking

- What weather might close your school?

Tornado

Sandstorm

Hurricane

Flood

The seasons

There is a pattern of seasons during the year. In winter we stay inside to keep warm. In summer we like to be outside.

Lazy Rufus

Rufus Rabbit was born in the spring. Outside his burrow he found sweet, juicy grass. By summer he felt big and strong.

'Come and help us repair the burrows,' said the other rabbits.

'No!' said Rufus, 'I can do what I like. It's sunny and there is plenty of food. I'll sleep outside.'

In autumn, it rained. Rufus had to sneak carrots from a garden and sleep under some old hay.

Soon it began to snow. 'Oh! How cold it is in winter,' he moaned.

'I was silly,' he told the other rabbits. 'If you let me back into the burrow, I promise to work really hard next year.'

Do you believe him?

Why did Rufus Rabbit enjoy the spring and summer more than the autumn and winter?

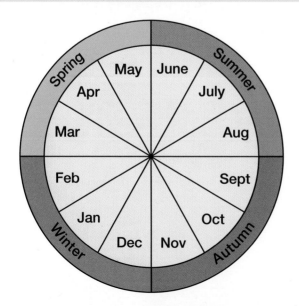

Talking
- What is the pattern of the seasons?

Spring

Summer

Autumn

Winter

Going round the sun

The seasons change as the Earth goes round the sun.
The sun is high in the sky in summer and low in winter.

The kittens can't agree

The kittens were arguing as usual.

'I don't like winter,' said Tim. 'It's so cold.'

'Summer is best,' said Dan. 'Hot sun for me!'

'Why do seasons change anyway?' asked Millie.

'Grandad said Father Christmas changes the weather,' said Tim. 'It's cold if we are naughty and warm if we are good.'

'I think it's the trees,' said Lyn. 'When they are tired it's winter. When they wake up it's summer.'

'I think it's the sun,' said Dan. 'When its fires are bright it's summer. When they die down it's winter.'

'That doesn't seem right,' thought Millie. 'It's to do with the Earth going round the sun.'

They all began arguing again.

What do you think?

Which kitten had the best answer?

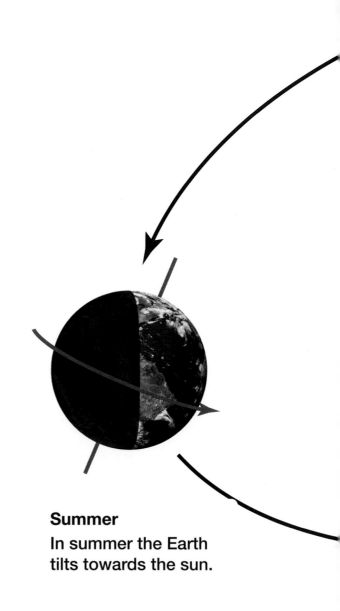

Summer
In summer the Earth tilts towards the sun.

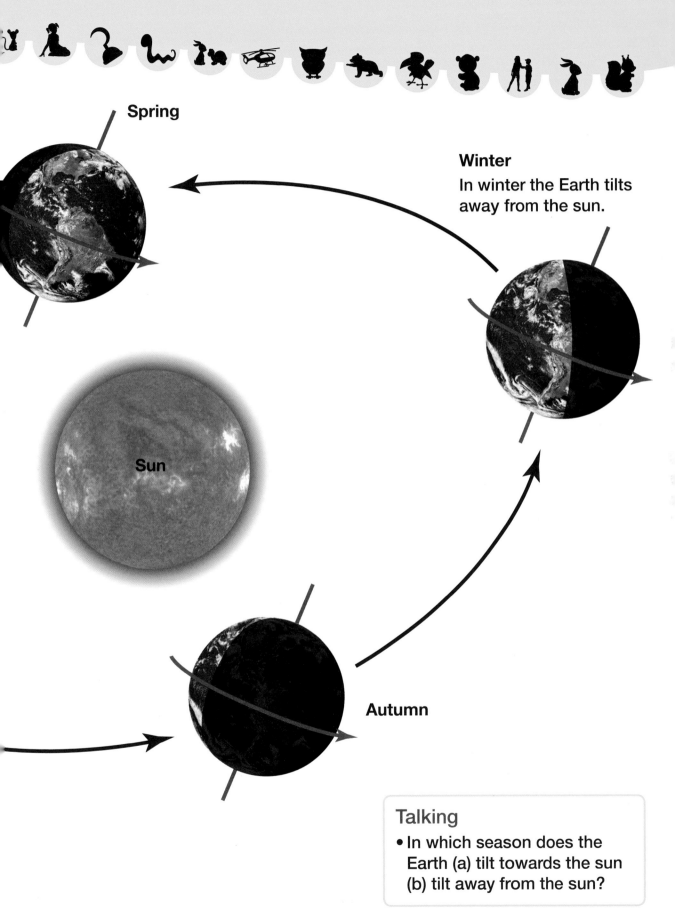

Talking
- In which season does the Earth (a) tilt towards the sun (b) tilt away from the sun?

Shelter

Houses give us shelter from the weather. They are where we feel safe and warm.

Camping in the garden

'We want to camp in the garden,' said Becky and Matt. 'No,' said their mother, 'You won't like it,'

'We will! We will!' answered the children. 'Well I did warn you,' replied mother.

That afternoon it was warm and sunny. The children put up their tent. They put sleeping bags, food and drink inside.

'It's our little home,' they said.

In the night the wind began to blow. The trees rustled. An owl hooted. 'I'm frightened,' said Becky. 'It's a bit cold,' said Matt.

In the morning mother found them indoors, safely tucked up in bed.

Why did Becky and Matt think their tent had become a little home?

Talking

- How do the items in the camping kit make the tent a home?

Camping Kit

- Cooking pots
- Tent
- Sleeping bag
- Water containers
- Table and chairs
- Games and toys

Houses around the world

People build houses in different ways. They can be made of brick, wood, stone or concrete. All houses give us shelter.

Town and country mouse

Bill Mouse lived in the country. His cousin Rita Mouse came from the city for a holiday.

'You will like it here,' said Bill. 'It's peaceful and quiet.'

They walked through the village. They passed thatched cottages and old wooden houses. At the end was a little caravan park with holiday homes. That was all.

After a week Rita said, 'It's very quiet. Come to the city with me.'

Off they went. Rita took Bill along her street. There were single houses, houses joined together and great blocks of flats. People and cars rushed everywhere. They were chased by a cat.

'I'm going back to the country,' said Bill. 'The city is too exciting for me!'

What types of houses did Bill and Rita Mouse see?

Talking

- Are the houses in the drawings all built of the same materials?

33

Living in a village

A village is a place where houses are grouped together. There are also schools, shops and other places which help people in their daily lives.

Max Mole loses his hat

Max Mole was feeling sad. He could not find his hat. He went from one end of the village to the other asking everyone, 'Did I leave my hat here?'

He asked in the shop, but they said, 'No.'

He asked at the garage, but they said, 'No.'

He asked at the church, but they said, 'No.'

He asked the people at the big house and in the cottages, but they all said, 'No.'

The last place in the village was the school.

'Have you seen my hat?' he asked.

'Yes,' they said, 'it's on your head!'

What places did Max Mole visit?

Shop

Cottages

Church

Pond

Primary school

Talking
- Why is a village more than just a group of houses?

Exploring local streets

There are lots of things to see in the streets around you. These are clues to the way we live and the things we need.

The fancy dress party

William was going to a fancy dress party dressed as a pirate. He had a pirate's hat, a red shirt, long boots and a cutlass made from cardboard and silver paper.

He ran quickly down the street. He bumped into a litter bin and tore his trousers.

He brushed against a newly painted bus stop and spoiled his shirt.

His hat was knocked off by a lamp post and squashed by a bus.

His boot caught in a drain and the heel broke. A dog ran away with his cutlass.

William was in a state when he arrived at the party. But he still won a prize. For being dressed as a scarecrow!

What was in William's way as he rushed down the street?

Talking
- Use the letters in the circles to help you work out what each photograph shows.

Under your feet

There are lots of pipes and wires under the pavement. They bring us water and gas from far away. Drains take away our waste.

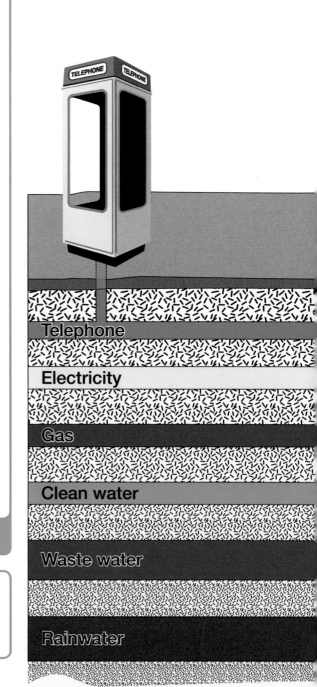

Winnie tries to sleep

Winnie Worm yawned. She was tired of making tunnels. She dozed sleepily. Suddenly the earth shook with a very loud noise. Winnie sat up as a large drill broke the roof.

'What's happening?' shouted Winnie. 'Sorry!' said a voice, 'I am putting in a new water pipe.'

Winnie settled down again. Brrrrrrrrrrrr! 'What is it?' she said. 'Sorry!' said a voice, 'I'm making a hole for a new gas pipe.'

Winnie tried to sleep again. Thud! Thud! Thud! 'What is it this time?' she complained.

'Sorry!' said a voice, 'I'm digging for a new electricity cable.' At last it was quiet. Bang! Bang! 'What now?' she moaned.

'Only me,' said Max Mole, 'Here are some earplugs.'

What disturbed Winnie Worm?

Talking

- Why do we need pipes and wires under the ground?

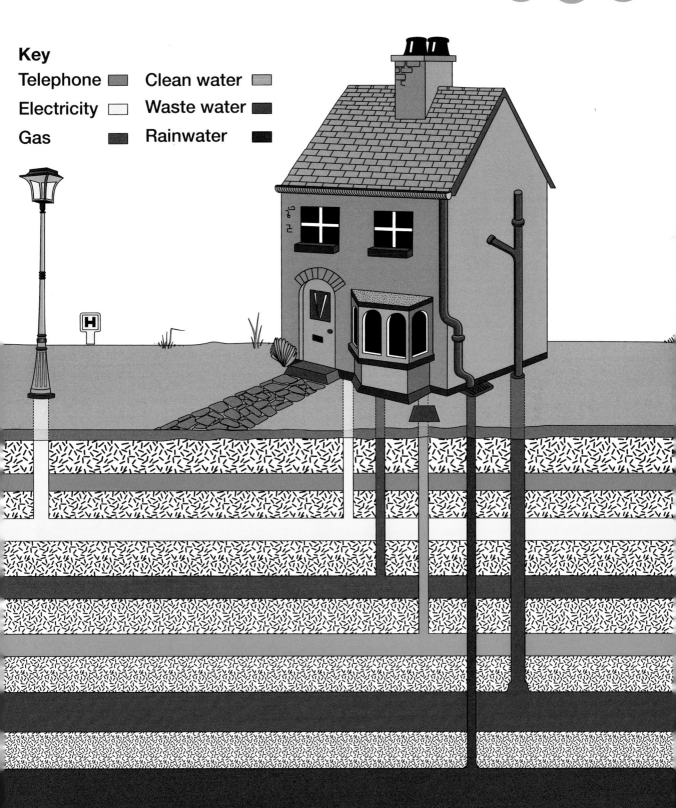

Maps and stories

Picture maps can show us about the places in songs and stories. They make it easier for us to see what happens.

The Hare and the Tortoise

The Hare was always boasting about how speedy he was. 'Hello, Hare,' said the Tortoise. 'I am going to eat lettuces on that hill.'

'You'll never get that far,' said Hare. 'I'll race you,' replied Tortoise.

Away dashed Hare, laughing. Soon he sat down and fell asleep by a tree. Tortoise plodded past.

Hare jumped up and rushed ahead. He came to a grassy bank. He sat down again and dozed off in the sun.

When he woke up there was no sign of Tortoise. Off he ran. At last he scrambled up the hill. Tortoise was already eating a lettuce leaf!

'How did you manage it?' panted Hare. 'I just kept plodding on,' answered Tortoise.

What places did Hare and Tortoise pass in their race?

Talking
- Talk about other stories and songs you know which are about places.

Treasure island

Maps show us where places are and how they are linked together. They help to stop us getting lost.

A lucky find

Tom and Karen were excited. Their father had bought a metal detector.

Sam, the boy next door, was jealous. He made a pretend treasure map to trick them.

'My grandad had this old map,' he said. 'See if you can find treasure with your detector.'

Tom and Karen excitedly took the map. They started off, going west from the village to the church. Nothing there. They went south to the lighthouse. Nothing there. They went east to the castle.

They were going to give up. Suddenly the metal detector beeped. They dug a hole and found an old Roman coin.

'We did find treasure,' they told Sam when they returned. 'It's going to the museum.' Sam wasn't laughing now.

What directions did Tom and Karen go in?

Talking
- What are the grid squares for the lighthouse, castle, church, mountain and village?

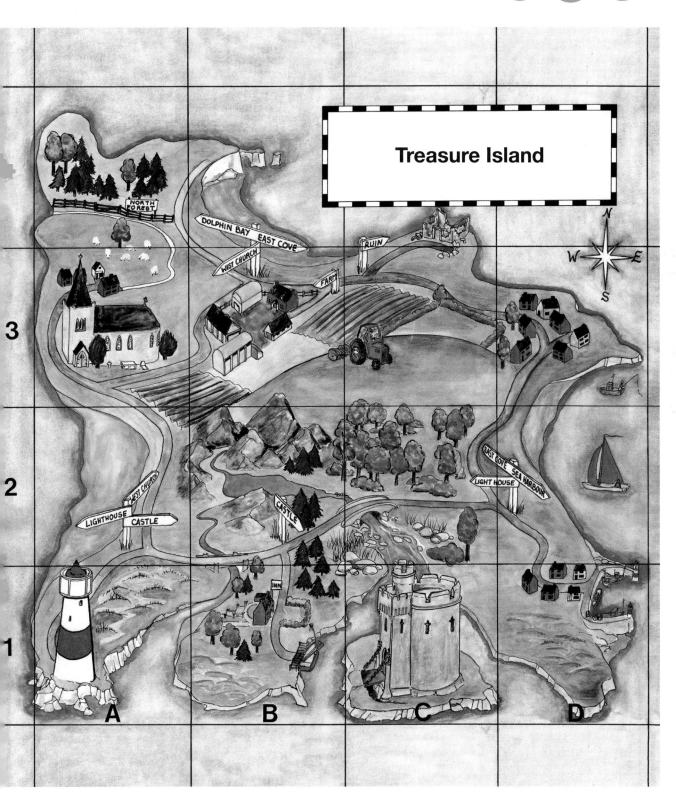

Different plans

Plans show us what places look like from above. This helps us to see their shape.

Talking
- Talk about the shape of some of the buildings and places that you know.

A new school

The children were excited. Their new school was opening.

Class 1, the youngest, went into their classroom first. It was just right.

Class 2 went in next. They could only just fit in their room.

Class 3, the eldest, were last. 'This is hopeless,' they complained. 'There is only space for half of us.'

The head teacher called the children into the hall. It was so small Class 1 filled it up!

'My office is too big,' said the head teacher. 'So is mine,' said the secretary.

The builder came, 'I expect you are pleased with your new school,' he said. 'It was quick to build because I decided to make all rooms the same size.'

'Boo!' shouted the children. 'What a bad idea!'

Why can't all rooms be the same size and shape?

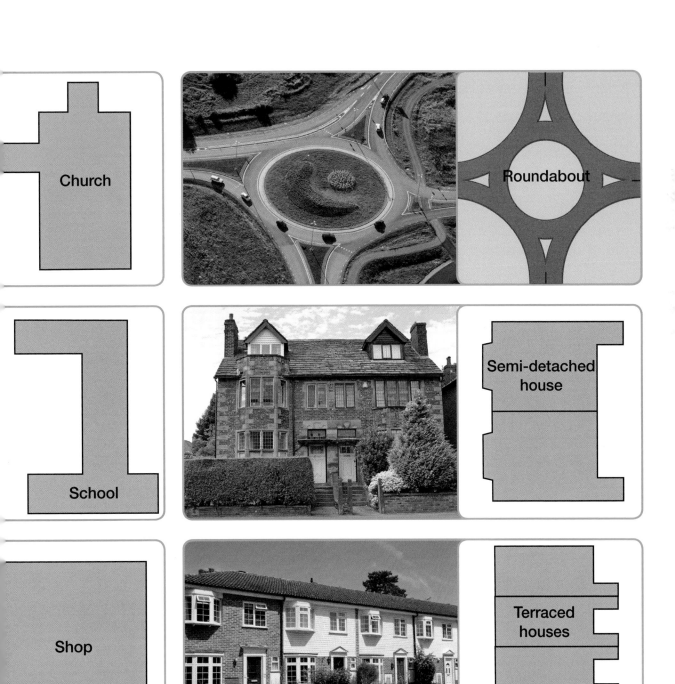

The view from above

Plans show what places look like from above. The higher we go in the sky the smaller places seem to be.

Hector Helicopter shows off

Hector Helicopter was showing off.

'I know what everything looks like,' he boasted. 'Because I fly so high, I can see things from above. I know the shapes of everything.'

'Look', said Hector, 'There is a football pitch. It's a rectangle shape. And do you see that circle? That's a roundabout.'

Suddenly Hector spotted a square.

'Good,' he said to himself. 'That square must be my landing pad. I will drop down and have a drink to keep me going.'

Down he came.

'Why have you landed here?' asked the pilot. 'We're on top of a block of flats miles from home.'

'Oh no!' groaned Hector. 'It looked just the right shape.'

Why did Hector Helicopter get confused?

Talking
- What different shapes can you see in the photograph?

46

UK countries

There are four countries in the UK. England is the largest. Northern Ireland is the smallest.

Edinburgh Castle, Scotland

Alfie tidies his room

'Please tidy up,' said Dad. Alfie Owl grumbled. 'Oh! Bother! I wish I wasn't here.' A wizard waved his magic wand.

Zing! Alfie was in a fir tree on a mountain.

'Where's this?' he asked. 'Scotland,' answered an eagle. 'It's cold and lonely,' shivered Alfie. 'Take me away.'

Zing! He was on rocks by a stormy sea.

'Where's this?' he asked. 'Northern Ireland on the Giant's Causeway,' answered a gull. 'Help!' cried Alfie, 'I don't like giants.'

Zing! He was on a grassy hillside surrounded by sheep.

'I'm hungry,' he said. 'Where's this?' 'Wales' answered a sheep. 'Have some grass.' 'Back to England,' wished Alfie.

Zing! He was home.

'Finished tidying?' asked Dad. 'Not quite,' said Alfie.

Which countries does Alfie visit?

Houses of Parliament, England

Talking

- Talk about the things you know about each UK country.

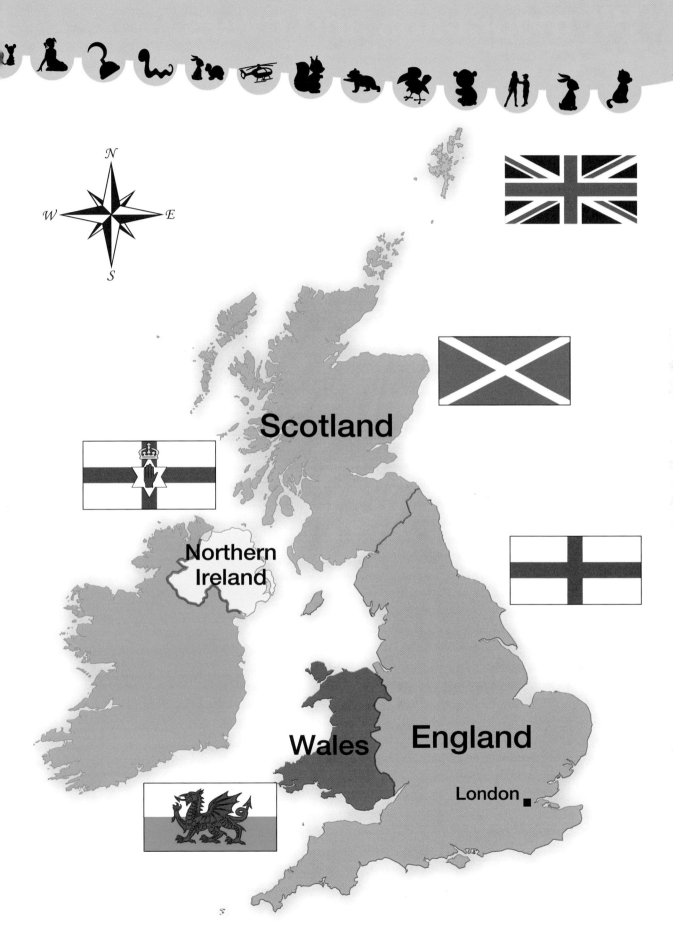

UK mountains and rivers

There are groups of mountains in the north and west of the UK. Rivers flow the flatter land in the south and east.

Mount Snowdon

Time for a story

Mrs Jones the teacher decided to tell the children a story. 'I'm going to tell you about a special place in the UK,' she said.

'Would you like to hear about the little train going up Mount Snowdon in a snowstorm?' 'No! No!' chorused the children.

'Would you like to hear about the journey of a baby eel swimming up the River Severn?' 'No! No!' cried the children.

'What about a holiday adventure on a boat in the Norfolk Broads?' 'No! No!' called the children.

'Well let's do some sums instead,' said Mrs Jones. 'No! No!' roared the children.

'Yes! Yes!' said Mrs Jones.

What three stories did Mrs Jones offer to tell the class?

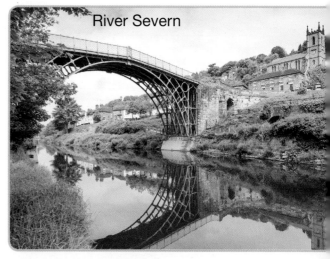

River Severn

Norfolk Broads

Talking
- What are the mountains, rivers and seas shown on the map?

Grampian Mountains
Southern Uplands
Lake District
Pennines
River Shannon
Mourne Mountains
Irish Sea
Mount Snowdon
Cambrian Mountains
River Severn
River Trent
Norfolk Broads
River Thames
North Sea
Atlantic Ocean
English Channel

51

Living in the arctic

The arctic is very cold and snowy. There are long, dark winters. Animals struggle to survive.

Finding the right place

The polar bear, reindeer and walrus were having a meeting.

The polar bear complained, 'I roam the snow all day. No one to talk to.'

The reindeer said, 'I wander about eating moss.'

'You're lucky,' said the walrus, 'Just fish, fish, fish for me.'

So they swapped places.

The polar bear dived into the water, the walrus flopped onto the moss and the reindeer swam out to sea.

The polar bear grew bored with swimming, the walrus didn't like moss and the reindeer tired of fish.

'I think we should stay where we belong,' said the polar bear and they all agreed. 'Still a change is as good as a rest,' said the reindeer.

Why was each animal complaining?

Polar bear

Reindeer

Walrus

Talking
- Do you think you would like to visit the arctic?

53

Living in the rainforest

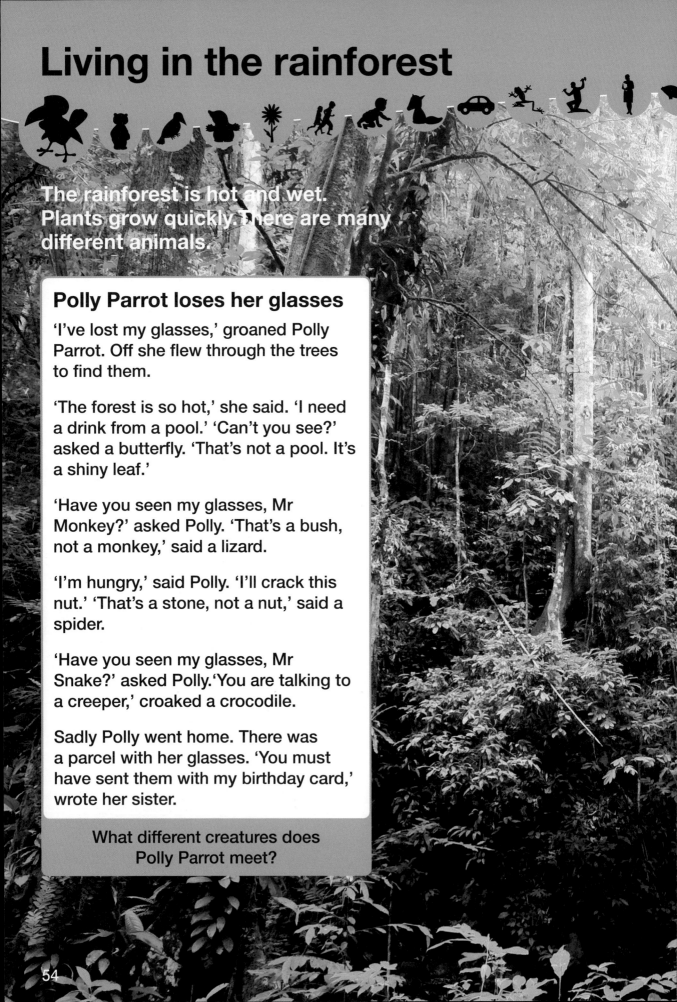

The rainforest is hot and wet. Plants grow quickly. There are many different animals.

Polly Parrot loses her glasses

'I've lost my glasses,' groaned Polly Parrot. Off she flew through the trees to find them.

'The forest is so hot,' she said. 'I need a drink from a pool.' 'Can't you see?' asked a butterfly. 'That's not a pool. It's a shiny leaf.'

'Have you seen my glasses, Mr Monkey?' asked Polly. 'That's a bush, not a monkey,' said a lizard.

'I'm hungry,' said Polly. 'I'll crack this nut.' 'That's a stone, not a nut,' said a spider.

'Have you seen my glasses, Mr Snake?' asked Polly. 'You are talking to a creeper,' croaked a crocodile.

Sadly Polly went home. There was a parcel with her glasses. 'You must have sent them with my birthday card,' wrote her sister.

What different creatures does Polly Parrot meet?

Talking
- Think of all the words you know which are about the rainforest.

Living in the desert

Most deserts are very hot and dry. Plants and creatures have to protect themselves from the heat.

Stranded in the desert

Kate Koala lived in Australia. She decided to explore the desert sights.

Her friends warned, 'If your car breaks down stay in the shade and drink plenty of water.'

The desert went on for thousands of miles. It was bigger than Kate ever imagined. She drove on and on. She ran out of petrol. 'What shall I do?' she sobbed.'

A kangaroo hopped by. 'Are you in trouble?' he asked.

'Yes,' said Kate. 'Please find help!' Off hopped the kangaroo. Later Kate heard a helicopter.

The pilot rushed over. 'Here is the water you wanted,' he said. 'I needed petrol' cried Kate. 'Never mind,' said the pilot. 'You can ride back with me.'

Why was Kate Koala in danger when her car ran out of petrol?

Talking
- Retell the story of Kate's adventure. Add information from the photographs.

Ayers Rock

Desert road

Water bottles

Eucalyptus trees

Helicopter

Animals around the world

It has taken millions of years for different plants and creatures to develop. We need to remember that we share the world with them.

Polar bear
Greenland

Eagle
Rocky Mountains

Butterfly
Amazon rainforest

King Penguin
Antarctica

Talking
- Which of the creatures live in (a) the rainforest, (b) the desert and (c) polar lands?

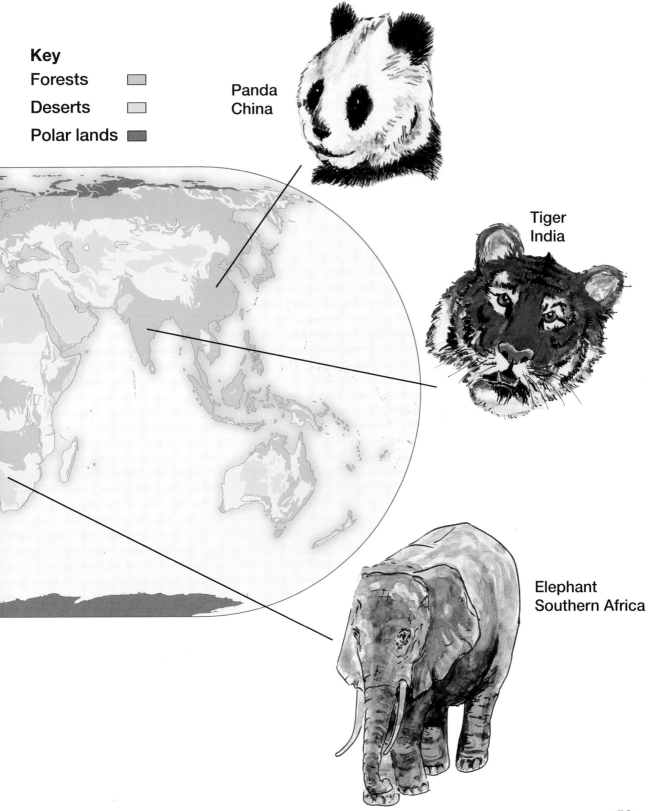

Key
Forests
Deserts
Polar lands

Panda
China

Tiger
India

Elephant
Southern Africa

World continents

ARCTIC

North America

ATLANTIC OCEAN

South America

PACIFIC OCEAN

Talking
- In what way is each continent different?

There are seven continents and five oceans.

OCEAN

rope

Asia

frica

PACIFIC OCEAN

INDIAN OCEAN

Oceania

ntarctica

World countries

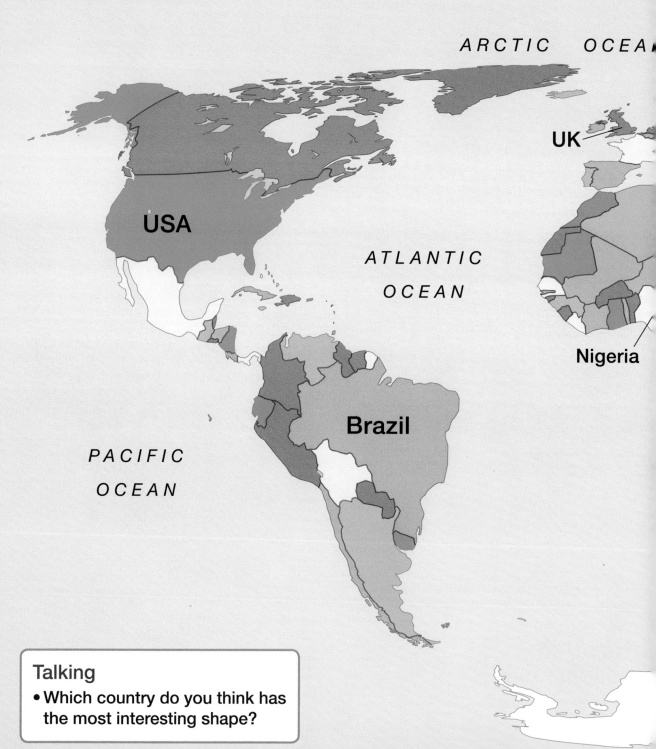

Talking
- Which country do you think has the most interesting shape?

There are about 200 countries in the world.

USA Brazil UK Nigeria China Australia

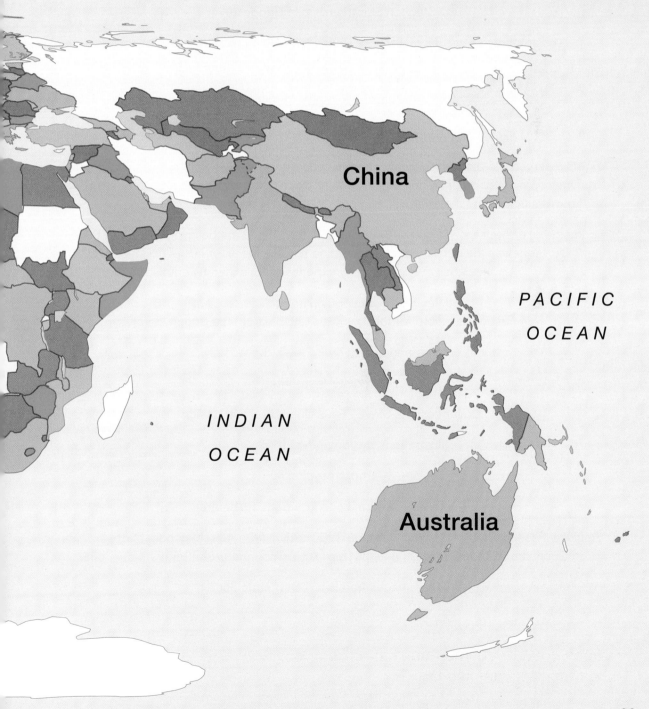

Primary Geography Book 1 and 2
Collins
An imprint of HarperCollins Publishers
Westerhill Road
Bishopbriggs
Glasgow
G64 2QT

© HarperCollins Publishers 2014
Maps © Collins Bartholomew 2014

First published 2014, reprinted 2014

© Stephen Scoffham, Colin Bridge 2014

The authors assert their moral right to be identified as the authors of this work.

ISBN 978-0-00-756358-6

Imp 004

Collins ® is a registered trademark of HarperCollins Publishers Ltd

All rights reserved. No part of this publication may be reproduced, stored in a retrieval system, or transmitted in any form or by any means, electronic, mechanical, photocopying, recording or otherwise, without the prior written permission of the publisher or copyright owners.

The contents of this edition of Primary Geography Book 1 and 2 are believed correct at the time of printing. Nevertheless the publishers can accept no responsibility for errors or omissions, changes in the detail given, or for any expense or loss thereby caused.

British Library Cataloguing in Publication Data
A catalogue record for this book is available from the British Library.

Printed and bound by L.E.G.O. S.p.A., Italy

Acknowledgements

Cover designs Steve Evans illustration and design

Illustrations by Jouve Pvt Ltd pp 38, 39

Photo credits:

(t = top b = bottom l = left r = right c = centre)

© Stanislav Fosenbauer/Shutterstock.com p57t; © pornvit_v/Shutterstock.com p25b; © Stephen Scoffham p20b, p21t, p21b, p40, p41t, p41b; © Tupungato/Shutterstock.com p35

All other images from www.shutterstock.com